Now We're Talking

The Story of Theodore W. Case and Sound-on-film

By

Antonia K. Colella
and
Luke P. Colella

ISBN: 1-4107-8770-2 (e-book)
ISBN: 1-4107-9515-2 (Paperback)

Library of Congress Control Number: 2003096003

This book is printed on acid-free paper.

Printed in the United States of America
Bloomington, IN

1stBooks – rev. 10/28/03

For Luke — my son, moon and stars.
Without you, this book would never have been written.
— *Mom*

For the people of Auburn, the students at
State Street School in Skaneateles, New York, and the
citizens of the world from Vietnam to America.
— *Luke*

About the Authors

After touring the birthplace of talking movies, 8-year-old Luke Colella asked to buy a book about the famous inventor who put sound on film. He was told a biography was never written. Inspired, mother and son began filling the gap and writing the untold story of Theodore W. Case.

Now 10, Luke enjoys music, collecting baseball cards, sports, family, friends and school. His mother, Antonia (Toni) Colella, is a reading teacher with degrees from Rosary Hill College, George Washington University and Syracuse University. She has written articles on education for newspapers and professional journals, and in 1999 created the New York State Newbery Quiz Bowl. Mother and son enjoy life in Auburn, New York.

Pictured are members of the St. Paul's School Scientific Association in 1908. Ted Case (left of center - 3rd row) was elected to the club's highest student office. [Collection of the Case Research Lab Museum]

Ea discamus in terris, quorum scientia
perseveret in coelis

"That we may learn on earth the knowledge that would
preserve us in the heavens"

— Motto of St. Paul's School

translated by Patrick C. Lewis

Movietone was the name the Fox-Case Corporation gave to all Fox productions made with sound. The truck's interior carried equipment to record and reproduce Movietone newsreels. [Collection of the Case Research Lab Museum]

Contents

Acknowledgments ix

Introduction... x

Forefathers: (From 1635).. xv

I. Auburn, New York: (Early 1900s) 3

II. Education: (1901-1913) 7

III. Early Discoveries: (1914-1916)... 13

IV. Anti-submarine Device: (1917-1919).. 19

V. Sound-on-film: (1920-1922)... 27

VI. A Partnership: (1923) 35

VII. A President is Heard: (1924-1925) 47

VIII. Worldwide Newsreels: (1926-1927).. 55

IX. Movie Moguls: (1928-1929) 65

X. The After Years: (1930-1944) 75

XI. Reflections.. 83

Case Research Lab Museum.. 85

Dates in the life of Ted Case: 87

In 1927 only Fox-Case Movietone (see logo on truck) had the sound capability to film people and events around the world. This news crew is interviewing June Collyer and George O'Brien. [Collection of the Case Research Lab Museum]

Acknowledgments

We wish to give our gratitude to the many people who offered their assistance on this project:

To Mrs. Jane Case Tuttle, whose lovely stories and warm encouragement enlightened our journey from beginning to end, and to Mr. Ted Case III, whose thoughts on the creative process (see Introduction) and comments on the style and content of the book were invaluable.

To the helpful and caring people at the Case Research Lab Museum, the Cayuga Museum of History and Art, the Seymour Library and History Room, the Cayuga Community College Library, the Cayuga County Historian's Office, Casowasco, Unity House, and the Alumni Offices of Yale University, St. Paul's School, and Manlius Pebble Hill School (formerly St. John's Military Academy). Without their generous assistance, the story would not be as thorough and complete.

To the many people who gave freely of their time and hearts to read our manuscript and to offer suggestions that were deeply appreciated.

To my brother, Luke's uncle and godfather, whose description of our first draft as "a quick and interesting read" inspired us to continue.

Above all, to my parents, Luke's grandparents. We could always depend on their love, encouragement, insightful comments, and wonderful family dinners. We love you both more than words can say.

T & L

Introduction

Theodore W. Case was my grandfather and this book is about him. It tells about his life and accomplishments but it tells much more. It tells more because through it we can all learn something about ourselves.

Theodore W. Case was a scientist and an inventor. He created the key inventions that made talking pictures possible along with many other important inventions. That creative nature lies at the heart of each of us. That creative nature isn't just one that creates things out in the world. It is a nature that creates a life — our life.

As you read Ms. Colella and Luke's story about my grandfather you will see three things that he did over and over. These things are the natural order of the creative process.

First, he gathered information. He wanted to know what had already been done in his area of interest. I have a letter from his father (my great grandfather) where he advises his son, "I advise you, as you should do in all cases, first find out what has been done." That gathering of information not only told him what had already been done, but also revealed all of the possibilities for creating something new.

Second, he created a picture in his mind of what he wanted — his goal or purpose. You will see an excerpt in this book of a letter from him to his mother where he says, "… my time now is taken up in experimenting with my selenium cell *with the idea in mind* of photographing sound waves." Picturing a goal is key to the creative process, and that miracle of our imagination is what makes human beings so unique. Without that *idea in mind,* there would have been no way to recognize what was useful.

And third, he experimented — he took action in a trial-and-error manner. In Chapter IV, Ms. Colella notes, "The first experiments tested the minerals and crystals found in Dr. Willard's cabinet. The men were looking for materials with extreme sensitivity to light." If something offered the possibility of moving forward he would try it. He did not insist that he would be right; he was curious about what was right. At the heart of his experimentation was the question: what works?

So enjoy the story of my grandfather. He invented some really great things and lived a very interesting life. At the same time, remember that each of us is also an inventor. Whether we are creating a new invention or creating a great day, we also follow that same creative process.

As you read, see how he gathered information. Look for examples of how he pictured his goal. Finally, notice how he experimented by trying those things that might be useful. These activities are the path of our creative nature — the path to what works. Finding out what works might lead you to a very important invention or a very exciting day. Good luck!

Theodore W. Case III

He would transform how the world sees and hears itself.

This photograph of Willard Erastus Case, Ted's father, was taken in the late 19th century. Ted would inherit his father's scientific intellect and passion for discovery. [Collection of the Case Research Lab Museum]

Forefathers:

(From 1635)

First in America

John Case sailed from England in 1635 aboard the ship Dorset. He helped start Simsbury, Connecticut, and the colony elected him its first constable. The Case family would remain in and around Connecticut for nearly 200 years.

Great Grandfather

Erastus Case was born in 1789. He moved to Auburn, New York, in 1843 when his daughter married Dr. Sylvester Willard, that city's prominent physician. Erastus was a pioneer in the railroad building of the country and a prudent businessman. With Dr. Willard, he gave generously to Auburn's Theological Seminary and First Presbyterian Church.

Grandfather

Theodore Pettibone Case was born in 1818. He did not follow his father in business, but studied Latin and Greek and the scientific discoveries of the day. He and his wife, Frances Fitch, displayed compassion for the struggles of their community. They contributed to the missions of the Presbyterian Church, requesting to remain anonymous.

Father

Willard Erastus Case was born in 1857. He graduated from law school, but was lured by the excitement of science. At the age of 29, Willard spoke before the Royal Society of London and demonstrated how heat energy converted to mechanical energy. He obtained numerous patents and associated with leading scientists around the world.

In 1904, he honored his parents by presenting the Case Memorial building to the Seymour Library Association of Auburn.

Willard married Eva Fidelia Caldwell. On December 12, 1888, a son was born who would inherit the family's scientific spirit and insatiable curiosity. He would transform how the world sees and hears itself. His name was ...

Theodore Willard Case.

Early on, he would climb upon a lab stool and watch his father measure, mix and record results of his studies.

Teddy Case, 3 years old. [Collection of the Case Research Lab Museum]

I. Auburn, New York:

(Early 1900s)

Auburn's Heyday

Auburn, New York, was a bustling community in the early 1900s. Called the cradle of industrial creativity, the city supported over 300 businesses and factories. Strong and diverse companies made carriages and carpets, shoes, sausages, fishing tackle, farm machinery, maps, rope, pianos, buttons and brass.

Auburn, too, was a center for popular entertainment. People from all over Central New York came by carriage and train to the many elegant and stately theaters. There was something for everyone: Shakespearean plays and vaudeville acts, orchestras and operas, performances by John Philip Sousa and programs by Mark Twain. Auburn was synonymous with the performing arts.

Perhaps less known but equally important, Auburn was celebrated as a seat of science. Willard E. Case established and equipped two of the finest private laboratories in the country — one at his home and the other at the family summer estate on nearby Owasco Lake. To these hubs of research came the leading scientists of the time. They talked about transportation, communication and industrialization, and they contemplated the far-reaching impact of electricity.

At Home

In a sense, Ted Case was always at home in the world of science. Early on, he would climb upon a lab stool and watch his father measure, mix and record results of his studies. He saw firsthand that science was less about rush and recognition and more about pursuing ideas with steadfast determination, perseverance and utmost integrity.

Ted would inherit his father's attitude toward scientific inquiry. The discipline would lead him to invent one of the great advances of the 20th century.

Dearest Mother,

Yesterday I at last succeeded in transmitting sound by light.

This photograph shows Ted Case setting up a camera in front of a mirror to take a self-portrait. He was fascinated by the possible uses of light waves. [Collection of the Case Research Lab Museum]

II. Education:

(1901-1913)

St. John's Military Academy

In September 1901, 13-year-old Ted went off to St. John's Military Academy, a boarding school in Manlius, New York. Like the cadets at West Point, these younger cadets wore heavy gray uniforms and followed a regimented code of conduct. Ted earned good grades and played shortstop on the junior baseball team. Yet, unhappiness haunted him. One spring day he ran away, much to the disappointment of his parents.

St. Paul's School

From 1903 to 1908, Ted attended St. Paul's School in Concord, New Hampshire. The school prepared students for Harvard, Princeton and Yale. The curriculum was quite challenging. In his senior year, Ted studied the New Testament, Latin, French, German, mathematics, history and English, and he read books by the great classical and contemporary writers.

In addition, Ted formally studied physics. The course consisted of class lectures and extensive work in the laboratory. Students worked with gravity, density, force, friction, light, heat and electricity. He was enthralled.

Ted Case (third row, second from left) played on the Old Hundred football team at St. Paul's School. [Collection of the Case Research Lab Museum]

8

Ted excelled in science throughout his high-school years and earned the respect of both faculty and friends. In his final year, he was elected to the highest student office in the St. Paul's School Scientific Association.

Yale University

In 1908, Ted entered Yale University. Like the majority of college students, he played outdoor sports, preferring football, baseball and soccer; nevertheless, science consumed most of his time.

Ted was especially fascinated with the properties of light and sound and the relationship between them. He became preoccupied with the possibility of recording sound by changing sound waves into light waves.

In 1910, on the top of a page in his lab notebook, Ted wrote the heading To do someday. Listed below were three theories he wanted to test. The tests involved light, magnetism and selenium, a chemical element capable of converting light directly into electricity. Ted studied selenium at length. On January 22, 1911, he wrote:

> Dearest Mother,
>   Well the prom went off finely and we all had a wonderful time. Most of my time now is taken up in experimenting with my selenium cell with the idea in mind of photographing sound waves...

9

Less than one month later, on February 12, Ted wrote about his success:

> Dearest Mother,
>
> Yesterday I at last succeeded in transmitting sound by light. The eye could not detect the variation of the light at all .... The reproduction of the voice was perfect. Next I have to set up an apparatus for very delicate photographing of the light variations. It is very interesting work and gives me something to do alright.
>
> The time has passed very quickly indeed. It won't be long now before I am back home for Easter Vacation.
>
> Lots of love
> Ted

After Graduation

Ted Case graduated from Yale University in 1912 with a Bachelor of Arts degree. He took up the study of law at Harvard University for six months, but did not like it. Physics and chemistry proved more compelling. After traveling for a year, he went home to Auburn to work with his father in research.

There is every evidence that he is blazing out

a new trail…

Yale alumnus, Ted Case, in his mid-20s. [Collection of the Case Research Lab Museum]

III. Early Discoveries:

(1914-1916)

Laboratory Notes

Ted Case recorded copiously in his large hardcover notebooks. On his 27th birthday, he filled three pages with experiments on electrodes and the effects of sunlight, solutions and water. Ted's laboratory was his life. It was quite common for him to work long hours and to make multiple entries, some titled *Evening* and *Later in Evening*. He had a flowing, handsome script and his writings read like personal conversations:

Seemed to get better results with …

Result nothing. Whole cell spoiled.

Next tried a closed circuit.

In constructing the above cell, my theory has been …

Spent today in corroborating earlier experiments.

Worked beautifully!

This is getting very interesting now.

A couple drops of H_2O spoils all the reactions.

This worked … did not work … worked well!

A Wonderful Discovery

Like his father and grandfather, Ted kept up with the scientific trends of the day. Thomas A. Edison and a score of well-known scientists were working on the problem of extracting electrical energy from coal. Ted tried a different perspective. He decided to bypass coal and to generate electrical energy directly from sunlight.

Ted went to work and created a battery that produced electricity simply by exposing it to sunshine. He wrote to his father, who was working in Florida. Mr. Case replied, "You have struck a most wonderful discovery I believe." Ted responded, "This reasoning has made me so nervous I have tried to write as fast as I think."

Blazing Out a New Trail

On June 14, 1916, Ted Case presented his discovery before the New York Electrical Society at its annual luncheon in New York City. He titled his paper "Preliminary Notes on a New Way of Converting Light into Electrical Energy." T. C. Martin, who presided at the meeting, made these introductory comments:

> ... Mr. Case is a young engineer not yet particularly well known to the world or to fame, but I think we shall hear from him later. There is every evidence that he is blazing out a new trail and is doing part of the experimental work of which the electrical industries at this stage of the tremendous development stand more in need than ever.

Ted's reputation as a scientist spread quickly across the country. On July 17, 1916, a letter arrived from Mr. W. Lewis from the state of Washington. The envelope bore the 2-cent stamp of the day and only a two-line address in which Case's name was misspelled:

Inventor Mr. Theodore W. Chase
Auburn New York State

The September 1916 issue of the *Electrical Experimenter* also acknowledged Ted's presentation. The article described his electrical current as weak, but emphasized "... a beginning has been made and a way has been shown us. It will pave the way for greater things to come."

A Mineral Collection

Caroline and Georgiana Willard were the daughters of Jane Frances Case and Dr. Sylvester Willard.* Upon the death of the last surviving daughter, as she had no heirs, the entire family estate was willed to Willard Case. The inheritance included a massive amount of money, a Greek Revival mansion at 203 Genesee Street in Auburn, and an extensive mineral collection that would greatly affect future scientific discoveries.

☙

*In 1894, to honor their parents, Caroline and Georgiana built the Willard Memorial Chapel on the grounds of the Auburn Theological Seminary. The Tiffany Glass and Decorating Company of New York handcrafted the chapel's interior. Today it is recognized as the only complete and unaltered Tiffany chapel still in its original location.

He worked nonstop with the Navy and soon produced a way to transmit the human voice along waves of flickering light.

The Thalofide Cell reacted quickly to invisible light rays. Ted Case first used the cell in a secret communication system during WWI.
[Collection of the Case Research Lab Museum]

IV. Anti-submarine Device:

(1917-1919)

The Thalofide Cell

Behind the inherited mansion, Willard Case converted a greenhouse into a scientific laboratory for his son. They named it the Case Research Laboratory and hired a Cornell University chemist named Earl I. Sponable to assist in their investigations.

The first experiments tested the minerals and crystals found in Dr. Willard's cabinet. The men were looking for materials with extreme sensitivity to light. They found 19 new compounds that changed resistance when illuminated.

The most promising discovery was a compound of thallium sulfide. This substance was highly responsive to infrared light, the invisible rays beyond the red portion of the observable rainbow. Before long, the Case Research Laboratory used thallium sulfide to invent a new type of photoelectric cell. The cell reacted quickly to small changes of light unseen by the human eye. Ted named it the Thalofide Cell.

World War 1

In 1914, the world witnessed the beginning of World War I. Three years into the war, German U-boats (submarines) sank Allied ships transporting war materials, supplies and troops to Europe. The United States entered the war in 1917. America quickly set a plan for victory, calling upon all vital resources, both military and civilian.

Ted responded to the appeal at once. He became an associate at the Naval Experimental Station at New London, Connecticut, where he focused on the crisis of over-water communication. With untiring efforts, he worked to design a secret signaling system for Allied ships at sea.

Ted knew the system needed to use infrared rays since visible light signals could be detected. His Thalofide Cell became the principal receiving element in the operation. From a range of 18 miles, the cell collected infrared rays sent by a ship's signalman. It then read the beams and produced audible tones. Within weeks, Auburn's son had devised a way for ships to send coded dashes and dots past German submarines.

Ted, however, was not at rest. His longtime passion for light and sound had found a meaningful purpose. He worked nonstop with the Navy and soon produced a way to transmit the human voice along waves of flickering light. Allied fleets employed the secret communication system immediately.

Dr. Lee DeForest

Ted first became acquainted with Dr. Lee DeForest in 1917. Dr. DeForest was an 1896 graduate of Yale University and a popular scientist of the time. He had the reputation of inventing the audion tube that amplified electrical signals and made the creation of radio possible.

Ted needed amplifiers for his naval operation. He contacted the DeForest Radio Telephone & Telegraph Co. in New York City. Shipments of the product, however, were marred with mistakes: instructions were missing, socket connectors were detached, the tubes simply failed to amplify. When Ted asked Dr. DeForest to repair or to replace the items, the former sent regrets and an occasional reimbursement.

On July 31, 1917, the supply of amplifiers came to an abrupt end. The United States District Court ruled that the DeForest audion tubes were an infringement on another inventor's patent.

Three years would pass before Ted and Dr. Deforest would meet again.

This is a 1918 photograph of Alice Gertrude Eldred standing beside the Case Research Laboratory in Auburn, New York. She and Ted would soon marry and raise four children. [Collection of the Case Research Lab Museum]

Top-secret Mission

From 1917-1918, the Case Research Laboratory in Auburn, New York, saw much activity. The lab hired 10 people to produce Thalofide Cells for the Navy's top-secret mission. Employees were required to sign a legal document saying they would not divulge, not even to their families, anything seen, heard, or invented while employed by the Case Research Laboratory. The document continued:

> Work is of a secret nature and would do grievous injury to the United States Government and its Allies in the present war if it should become known. Doing so would constitute a form of treason to the United States Government.

One member of the staff was an attractive and intelligent young Auburn woman who carefully managed and cleaned the test tubes. Her name was Alice Gertrude Eldred. On November 26, 1918, 15 days after the war ended, Ted and Alice were married. Unfortunately, Ted's father would not live to see his son's wedding. Two weeks before the ceremony, Willard Case died during the great influenza epidemic.

Ted inherited the grand estate. He expanded the scientific workshop to include portions of the nearby carriage house.

Recognition

Ted's wartime accomplishments received recognition from academic institutions and from the United States government.

George Washington University bestowed upon him an honorary Master of Science degree. St. John's Military Academy honored him with the title of Commander of the Order of the Phoenix. The government's letter of appreciation follows:

Navy Department - Washington

May 19, 1919

Dear Sir:

The work carried on during the war by the Navy Department in developing anti-submarine devices and equipping vessels for anti-submarine operations had an important effect in restricting enemy submarine activities.

This result was made possible by the splendid assistance and co-operation of the many distinguished scientists, engineers and business men who were in one way or another associated with the Special Board on Anti - Submarine Devices, which had been appointed by the Department to supervise work of this nature.

The Navy Department wishes to express its appreciation of the valuable assistance rendered by you in this connection.

Yours very truly,
Franklin D. Roosevelt
Acting Secretary of the Navy.

To: Theodore W. Case of Case Research Laboratory, Auburn, N.Y.

With this process of transferring sound waves to light waves and back again to sound waves, his idea of a sound track became a reality.

The AEO Light was extremely sensitive to sound waves. The light recorded tiny lines of sound on motion picture film so that words, music and action could coincide. [Collection of the Case Research Lab Museum]

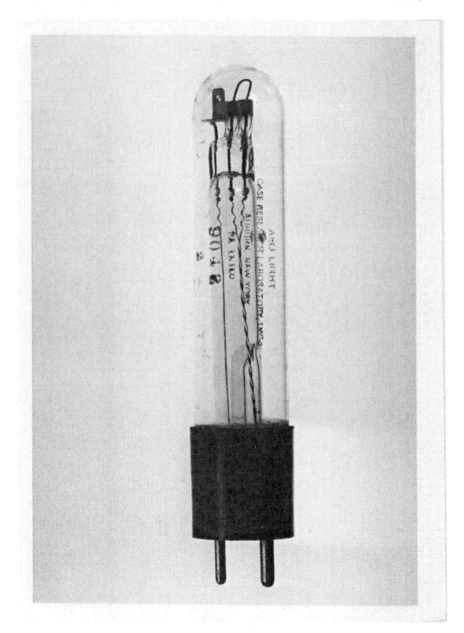

V. Sound-on-film:

(1920-1922)

Yale Questionnaire

After the war, Yale University sent out a questionnaire:

How did the war affect your life?
What is your life work?
What are your points of view, your hobbies?

Ted wrote about his Thalofide Cell and the methods he developed with the Navy that made signals audible. He still was not at liberty to say how the government used these inventions, only that the end of the war halted all further military development in voice transmission apparatus.

Next, Ted listed 10 titles of articles he had published in scientific journals. Then he wrote about his current work. He was researching the action of light on molecular and atomic structures. He was developing a new photoelectric cell that measured daylight as the eye saw it. The new cell reacted instantly to light variations and could produce enough electrical current to operate an automatic recorder. Ted said the cell was nearly complete. He hoped it would be useful to agriculture, electric lighting companies and weather bureaus.

Ted wrote about his political allegiance, saying he was always for the best man. About his personal life, he announced he was the proud father of two children, Theodore, Jr. and Barbara, and responded openly about his interests:

> My greatest hobby is my wife and my children and next is Research work after which hunting and fishing – then Radiophone Broadcasting!!

Near the bottom of the last page, Ted noted his unease while filling out the questionnaire. He said he felt like one of those old-time advertising mediums seen walking the streets of New York with a sign on the front and back.

International Race

Beginning in 1880, inventors from around the world presented proposals to photograph sound on motion picture film. Their descriptions, however, were too academic and ambiguous. Duplication was difficult. Systems failed. It was evident from all aspects that the necessary tools to couple sound and moving picture were not yet available.

One participant in this international race for sound film was Dr. Lee DeForest. In 1920, DeForest contacted the Case Research Laboratory to purchase Ted's Thalofide Cell, hearing how well it reproduced sound. Ted was excited. He welcomed the opportunity to work with Dr. DeForest and agreed to sell him his cells.

In 1922, Ted received a second request from Dr. DeForest. The New York scientist was trying to make a recording of his voice with tungsten filament lamps, but without much success. He could only reach the stage of determining that his words were not running backward through the projector. Could Ted solve the problem of recording sound on film?

The AEO Light

Unknown to the public, Ted had already discovered a way to record voices. Since the war, he worked to perfect the recording system's main tool, a gas discharge light. The light was 1 inch in diameter by 5 inches long and had a filament coated with alkaline earth oxide (AEO light).

Ted discovered that the AEO light was so sensitive to amplified vibrations that it could catch sound and transform it into light. The light could reflect upon motion picture film and register there at the same time as the action.

Ted attached the AEO light to the back of a specially constructed motion picture camera. The photographed sound waves appeared on the film as a narrow strip of wavy light and dark lines. The lines ran along the left side of the film, between the image and the sprocket holes.

Within the month, Ted further developed the AEO light that recorded sound and paired it with the Thalofide Cell that reproduced sound. With this process of transferring sound waves to light waves and back again to sound waves, his idea of a sound track became a reality.

On December 14, 1922, Ted wired Dr. DeForest at 318 East 48th Street in New York City.

> Can You take night train for Auburn?
> Important Development.
> Theodore W. Case

Five days later, DeForest replied with a request for Ted to send him the bulbs as soon as convenient. Ted wrote on December 22, 1922:

> Dear Dr. DeForest,
>
> I am sending down to you by the porter on tonight's sleeper the newest light which is very excellent. It may not look very bright to you but it is extremely rich in the violet. ...I am fixing up a new camera which ought to be ready some time next week. Let me know what you think of this last light after you have tried it.
>
> Very truly,
> CASE RESEARCH LABORATORY

After examining the light, Dr. DeForest responded promptly by Western Union telegram:

NEW CELL GIVES NO SCRATCHES
WHATEVER IS GREAT IMPROVEMENT
OVER THE OTHERS BY ALL MEANS
ARRANGE TO MAKE THIS TYPE. DEFOREST
TEL & TEL CO.

My dear Mr. Edison,

In this drawing of a remodeled Bell and Howell camera for sound recording, one can see Ted Case's AEO light. [Collection of the Case Research Lab Museum]

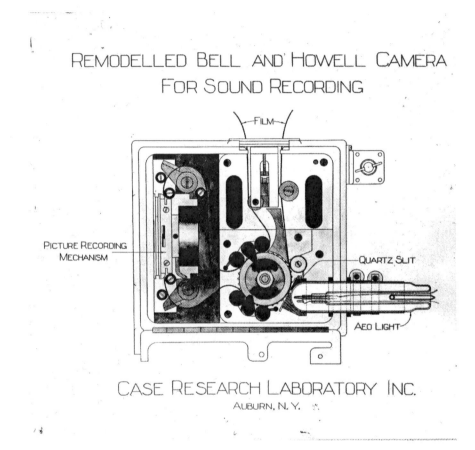

REMODELLED BELL AND HOWELL CAMERA
FOR SOUND RECORDING

CASE RESEARCH LABORATORY INC.
AUBURN, N. Y.

VI. A Partnership:

(1923)

Thomas A. Edison

Ted Case loved talking about science. Always interested in taking an invention one step further, he enjoyed hypothesizing with other scientists and extending to them an open invitation to visit his laboratory in beautiful Auburn.

On January 29, 1923, Ted wrote a letter to Thomas A. Edison. Mr. Edison was the inventor of the incandescent light bulb and the phonograph.

Mr. Thomas A. Edison   Orange, N.J.

My dear Mr. Edison,

A few weeks ago I happened to run into a strange phenomena in connection with the reproduction of music from one of your phonograph records. If there should be by any chance one of your research men in this part of the country, I would be pleased to have him stop here and we can see if it would be of practical utility. It has to do with the reproduction of our records by electrical instead of vibrational means and eliminates scratching noises.

Very truly,
Case Research Laboratory

35

On February 1, 1923, Edison's private secretary cabled Ted with a reply and a request:

Mr. T. W. Case,
Case Research Laboratory,
Auburn, New York

Dear Sir:

Your letter of January 29th was brought to Mr. Edison's personal attention. He wishes me to say that our research men stay in the Laboratory and do not travel. He also says there are a number of stunts for reproducers but none so good as the present system. If scratching noises are eliminated, then the fine overtones will also be eliminated. The trouble is that loud music is what the people want at present.

How about your photocell? Have you any printed matter that you can send Mr. Edison in regard to it.

Yours very truly,
W. H. Meadowcroft
Assistant to Mr. Edison.

This is a letter to Ted Case from Thomas A. Edison.
[Courtesy of Theodore W. Case III]

From the Laboratory

of

Thomas A. Edison,

Orange, N.J.

February 6, 1923

Mr. T. W. Case,
Case Research Laboratory,
Auburn, N.Y.

Dear Mr. Case:

Allow me to thank you for your letter of Feb. 3rd with its enclosures, which you sent to Mr. Meadowcroft, my Assistant.

Do you make a cell, which, when daylight appears, will actuate a relay to open a circuit connected to an electric lamp? I think there is a demand for this if crude cheap devices could be made practical.

There is a device made of metal, which is used on the flashlight posts at road crossings, but they are not cheap. The coming of daylight shuts off the current, and when it is dark or semi-dark the circuit is closed and starts the light.

Yours very truly,

Thos A Edison

TAE:FTR

Without delay, Ted sent information on his photoelectric cell. Mr. Edison replied personally on February 6, 1923. He thanked Ted for his enclosures and asked Ted: "Do you make a cell, which, when daylight appears, will actuate a relay to open a circuit connected to an electric lamp?"

Ted's response on February 8, 1923 informed Mr. Edison that he was under an agreement with the Leeds and Northrup Company to license the commercial use of his cells only with their apparatus. "If you are sufficiently interested," Ted wrote, "you might write to Mr. L. O. Heath of that firm in respect to this apparatus as they are entirely familiar with my cell."

Awestruck

On February 23, 1923, Ted Case and Earl Sponable carried their new camera to Dr. DeForest's studio. For the first time, Dr. DeForest saw and heard the synchronization of moving pictures with sound. He was awestruck and immediately discussed forming a company. Ted conferred with his attorney, John Taber of Auburn. Mr. Taber advised Ted to proceed with caution. He knew of Dr. DeForest's failed companies and fraudulent patents.

On March 17, 1923, Ted received a letter from Dr. DeForest. It suggested the two men "get together without interference of their attorneys." Dr. DeForest pleaded, "I have been working like the devil for over three years, building a market for the Thalofide Cell."

He promised that Ted would receive full credit for his inventions in all press promotions.

Ted was not able to act immediately on Dr. DeForest's offer. His family life posed pressing issues. Their second daughter had arrived and the three children, Ted, Jr., Barbara, and Jane Frances, greatly influenced his attention to work and travel. On April 17, 1923, Ted sent a wire to Dr. DeForest:

BARBARA VERY SICK. TAKE UP QUESTION WITH TABER ANYTIME YOU DESIRE. TW CASE

One week later, Mr. Taber approved a partnership between the DeForest Phonofilm Company and the Case Research Laboratory. Dr. DeForest gained a commercial license to use AEO lights and Thalofide Cells. Mr. Taber, however, retained the right to intervene and to assist in the protection of Ted's patents.

These musicians played for one of Ted Case's early tests of sound film. Notice the lines of photographed sound waves on the left side of the filmstrip. [Collection of the Case Research Lab Museum]

Newspaper Headlines

Ted's combination movies were making newspaper headlines. On March 17, 1923, the *Advertiser-Journal* ran the story: "Talking Pictures Invented by Auburn Man."

The article tells how Ted invited Auburnians to the carriage house to view his sound movies. The guests listened to violin and piano music played by students from the Eastman School of Music. They also heard a short recorded talk by Ted "whose voice was clear and without scratching or blurring sounds." On the film, Ted explained the sound-on-film process to the in-house audience:

> My voice to which you are now listening is recorded photographically on the standard motion picture film which you see running in the dark room behind you. The record is obtained on the film through a small slit two-thousandths of an inch in width by means of the small Heliolight. My voice is picked up by a microphone, then transformed into electrical vibrations, amplified, and impressed on to the glow light which made the photographic record. The film was then developed and is now being run through the machine. A constant light is projected through the slit and film record, where it then falls on the Thalofide cell which changes the light variations into electrical variations which are now being amplified and impressed onto the horn into which you are vainly looking at me.

41

The reporter admitted that he needed to quote Ted, as the scientific process was much too complicated for him to describe.

A Gentleman's Letter

By October 1923, 25 phonofilms were ready for exhibition. Dr. DeForest told Ted, "One can understand every word first time through," the AEO light was "working fine," and the new thermophone, a microphone Ted had recently invented, was "wonderful."

Ted was thus surprised to read a statement in the *New York Times* attributed to Dr. DeForest. On December 31, 1923, Ted composed a gentleman's letter:

Dear Dr. DeForest,

I see by the New York Times an article reported to have been given out by you. Part of the article reads as follows:

"The thermophone was perfected to a great extent by Theodore W. Case at his research lab at Auburn, N. Y., although the original idea was Dr. DeForest's."

May I inquire whether you were correctly quoted in the above quotation.

Truly,
Ted Case

42

Dr. DeForest admitted to making the comment. He tried to convince Ted that the public must not know anything about the identity of the true inventors for fear of others stealing their technology. Ted was unconvinced he had heard the truth. He continued to have his patent attorneys verify all descriptions of his scientific developments.

For the first time in history, the gestures and voice of an American president appeared on a motion picture screen.

The 1925 film of Gus Visser and his singing duck was honored by the Library of Congress in 2002. [Collection of the Case Research Lab Museum]

VII. A President is Heard:

(1924-1925)

Phonofilm "shorts"

Ted's discovery of a way to make movies talk and laugh and sing was starting to transform American entertainment. In the days of silent movies, written dialogue appeared on the screen during breaks in the picture's action. The titles spoke for the actors. Now actors were speaking for themselves.

Not everyone, however, was in favor of talking movies. Some thought the intrusion of talk was vulgar. Silent movie directors such as D. W. Griffith told reporters they would never have the human voice in their movies. Silent movie stars with untrained voices dismissed the idea as foolish. An Associated Press article (September 3, 1923) stated that even Jack Warner of Warner Brothers Pictures, Inc. declared talking pictures would never prevail. Yet thousands were flocking to theaters to see and hear the new sensation — phonofilms.

Phonofilm "shorts" ran under 10 minutes and played before the silent movie. Eddie Cantor and a host of new talents were astounding audiences with their audible acts. Theaters promoted talking pictures as the greatest invention of the age, the *8th Wonder of the World!*

President Calvin Coolidge

Talking movies also changed the way Americans perceived their democracy. In July 1924, the committee to re-elect Calvin Coolidge contacted the DeForest Phonofilm Company. The president could not make a New York engagement and wanted to send a speech on film. Dr. DeForest, Mr. Sponable, and Ted eagerly agreed to travel to the nation's capital.

Republican campaign managers asked Ted to give the event his personal attention. Ted immediately designed a special attachment for open-air recording. On the day of the filming, he worked closely with the crew, approving the site and making sure his equipment operated perfectly.

Shortly before 2 o'clock, on a hot August afternoon, the president appeared on the south portico of the White House. Mrs. Coolidge, their son, John, and a number of Secret Service men followed. Ted saw that President Coolidge was tense and tried to put him at ease. He calmly explained to the president how the camera functioned. He even traded places to show President Coolidge how he would appear in the viewfinder.

With speech in hand, the president began: "The march of invention is faster in our days than ever before." President Coolidge spoke confidently and much longer than anticipated.

The election committee wanted the film developed at once and shown that night. When Ted inspected the projection area in the East Room of the White House, he felt the echoes in the large room would cause distortion in the reproduction of any film. The committee heeded Ted's advice and agreed to have the historic film returned to the Phonofilm studio.

No Credit for Case

Weeks later, a large dinner was held at the Friars Club in New York City. Among the guests were over 500 editors and publishers from every part of the United States and from points as far as Australia and Berlin. President Coolidge, though more than 200 miles away, was the main speaker.

After dinner, the master of ceremonies introduced the president. The lights went out and the screen came alive. For the first time in history, the gestures and voice of an American president appeared on a motion picture screen. Even President Coolidge's New England accent came through loud and clear. The audience was stunned.

Dr. DeForest attended the dinner. Wasting no time, he rushed to call a press conference and unabashedly claimed full credit for the success of the presidential film. Because of Dr. DeForest's self-promotion strategies, newspapers around the world that covered the breakthrough failed to acknowledge the man truly responsible for the evening's exceptional event — Ted Case.

By late September 1925, the Coolidge film and other disagreements brought the working arrangement between Ted and Dr. DeForest to an acrimonious end.

The New Attachment

Ted removed his equipment from Dr. DeForest's studio in New York City. In order to prevent Dr. DeForest from playing prerecorded phonofilms, he designed a new sound reproducing attachment.

In the new attachment, Ted moved the film soundtrack to 14.5 inches or 20 frames in front of the picture. Having a precise location of the sound in relation to the picture created a major advance in sound movie production. The distance allowed technicians to cut and splice film without disturbing the illusion and naturalness of the scene. This latest invention, which Ted moved below the camera head, introduced the process of sound editing to the movie industry.

People Close to Home

Ted appreciated the people close to home. He had a talent for teaching. On the evening of March 31, 1925, he invited members of the Cayuga County Radio Club to his carriage house studio. Ted explained the scientific differences between radio and phonofilm and answered numerous questions before showing a variety of his talking pictures. He even said he would repeat the session the following night for those members of the club who were unable to attend.

Three months later in June, Auburn held an "Exposition of Progress" on the grounds of the Auburn Theological Seminary. The industrial fair proudly displayed local inventions and innovations. The Case Research

Laboratory erected a tent in which Ted set up a camera to show his films. Enthusiastic crowds filled the makeshift screening area. They took special pleasure in a hilarious five-minute film of Gus Visser and his singing duck. Every time Visser sang "Ma He's Making Eyes at Me," the duck quacked — and the audience roared.*

On the following day, the *Advertiser-Journal* reported that Ted Case contributed to the scientific world something of which any community would boast with pride. "Case's sound movies," it said, "were one of the greatest achievements in the history of the city, and perhaps in the world."

☙

* *The clip of Theodore W. Case Sound Test: "Gus Visser and his Singing Duck" was shown on television during the Academy Awards in 1992. In 2002, The Library of Congress recognized its technical innovation and placed the film on the National Film Registry.*

At this private showing, Mr. Fox realized that of all the sound units on the market, Ted's optical sound system was the most advanced.

The man in the center with a mustache is William Fox.
[Collection of the Case Research Lab Museum]

VIII. Worldwide Newsreels:

(1926-1927)

An Open Field

Ted decided to continue working in talking movies. Many technical problems remained unsolved.

He conferred with the best patent lawyers in New York City. It was their opinion that the field of talking movies was wide open. No one yet seemed to have any fundamental patents on a sound-on-film system.

In March 1926, a college friend of Mr. Sponable's stopped by the Case laboratory. He worked for the Fox Film Corporation and suggested that Ted and Mr. Sponable bring their equipment to New York City and demonstrate for the Fox people.

When the company's president, William Fox, saw the film of a singing canary, he was highly suspicious. He thought the movie was a matter of trickery or ventriloquism. He had never seen synchronized music and action on one film, and he insisted that Ted reshow the film at his home. At this private showing, Mr. Fox realized that of all the sound units on the market, Ted's optical sound system was the most advanced.

On July 23, 1926, Mr. Fox and Ted reached an agreement. They formed the Fox-Case Corporation to which Ted turned over all rights in his talking picture technology. Ted also agreed to continue working in his Auburn laboratory. He would make recording lights and photoelectric cells, and continue research and development in synchronized films.

Movietone

The Fox-Case Corporation chose the name Movietone for all Fox productions that featured sound. They built a new studio at 460 West 54th Street and equipped it with the world's first sound stage for motion pictures. Ted invited many well-known entertainers to come to the studio for taping. He conducted various tests to determine the best uses for sound and to learn its limitations.

On January 21, 1927, at the Sam Harris Theater in New York City, Movietone played its first one-reel short film production. Like appetizers before the main course, the short played in conjunction with the premiere of the silent feature film *What Price Glory?*.

The *Evening Graphic* reviewed the debut:

> Movietone is an improvement on anything of its kind yet attempted. It records the voice more perfectly than any other system and is the result of experimentation by Theodore W. Case, who evolved the method of photographing sound waves on a strip of motion picture film.

Simplicity

Ted continued to develop his technology and its applications. In 1927, people got their news by talking face to face or on the telephone, by reading periodicals, listening to radio broadcasts, or going to biweekly silent newsreels. Ted had an idea. Movietone could give people a fresh, first-hand way to experience the world around them.

Ted wrote an article in the *Yale Scientific Magazine* titled "New Advances Made in Talking Movies." Here he alluded to his pioneering proposal:

> It has been the intention in building up the Movietone system to make it as simple as possible ... to put the sound record on the film in the same camera which takes the picture ...
>
> The resulting Movietone system is now so simple that portable sets have been built which may be placed in an automobile, and sound pictures may be taken almost anywhere. It is evident from this that the system is not limited to the studio.

The idea took form in early May 1927. Movietone produced the first sound newsreel filmed outside studio walls. The new "interest short" premiered at the Roxy Theater in New York City. It covered the United States Military Academy's 125th Anniversary parade. Spectators saw and heard the clicking of heels and the transfer of weapons as the cadets marched to the beat of the West Point Band.

A Washington newspaper reported that the film was overwhelming in both image and sound. The reproduction, it said, was virtually perfect.

Charles Lindbergh

On May 20, 1927, Fox-Case Movietone recorded the first sensation in sound film history. At 7:52 a.m., Charles Lindbergh flew into the skies of Long Island in a single-engine monoplane. The American aviator and engineer would accomplish the first nonstop solo flight between New York and Paris, France.

That very afternoon, eager crowds in packed New York City cinemas watched, and heard, the epic event. They wanted a chance to share in the thrill of the moment, to hear the full throttle of the engine charge down the runway. *Variety* magazine reported that as the Spirit of St. Louis came into view and then quickly disappeared in the morning mist, viewers were unable to contain their excitement. People, young and old, yelled and stomped. It were as if they had personally witnessed the dramatic departure at Roosevelt Field.

The following week, Movietone went to Washington, D.C., to film the bands playing and President Coolidge giving Lindbergh a hero's welcome home. The complete Lindbergh footage, from takeoff to tribute, was shown over and over and was sold out for weeks.

The box office was talking. By October 1927, Movietone News, the first commercially successful sound film, appeared all over New York. Display ads began listing the theaters in Manhattan and Brooklyn where the news series was showing. Talking newsreels were causing the full-length silent features to fall by the wayside.

Vitaphone

The industry was now flooded with a variety of sound systems. Movietone's top competitor was Warner Brothers Pictures, Inc. In 1927 Jack Warner and his brothers produced a motion picture with synchronized sound parts. The film was a musical, *The Jazz Singer*, starring Al Jolson.

Warner Brothers used a different sort of recording system, called Vitaphone. While Movietone photographed sound on film, Vitaphone recorded sound on a wax disk. To achieve synchronization, the disk played on a phonograph while a projector rolled the film.

The Vitaphone process had many problems. Warping and scratches on the disk were common complaints. Editing was impossible. Coordination was prone to trouble. A projectionist had to be extremely accurate and place the stylus on the exact starting point. If only one groove on the disk was missed, the movie was out of sync.

In one Vitaphone movie when a man opened his mouth to sing to his dreamy-eyed girlfriend, strums of a banjo unexpectedly filled the air. The audience doubled with laughter or yelled angrily and threw things at the screen.

Movietone News traveled the world. In this photograph, a Movietone truck is in front of the Colosseum, where combats between lions and Christians once entertained the emperors of ancient Rome. [Collection of the Case Research Lab Museum]

Still another snag with Vitaphone was its equipment. Coordinating two separately motored machines to achieve synchronization was quite cumbersome. By necessity, the only place Vitaphone pictures could be filmed was in a studio.

No Competition

The opinion in the field was unanimous. There was no competition. Fox-Case Movietone News had the lead.

Only Movietone News was capable of filming famous people and historic events on location around the world. It spanned the globe and recorded speeches by Benito Mussolini, lectures by George Bernard Shaw, interviews with the Archbishop of Canterbury, performances by the Vatican Choir, the changing of the guard at Buckingham Palace, tours of Rome's Colosseum and of Niagara Falls, and one of Ted's favorite events—the Army-Yale football game at the Yale Bowl.

Overnight, Fox-Case transformed the way people got their news and information. Ted's idea to simplify had been on the mark.

తా

Ted's technology had given a fresh dimension to the Western movie, when critics thought it to be at the end of its trail.

This photograph shows the Cisco Kid (left) in a stagecoach scene from *In Old Arizona.* Using the inventions of Ted Case, the 1928 Western was the first sound movie produced outdoors. [Collection of the Case Research Lab Museum]

IX. Movie Moguls:

(1928-1929)

Away from the Flurry

The Fox-Case Corporation took off in 1928. Technicians installed sound systems in more than 900 theaters at home and abroad. News units covered the globe from Mexico to the Middle East. Elaborate studios in New York, Hollywood and Europe produced full-length features with synchronized sound. Fox-Case studios had even made a cartoon with sound effects.

Away from the flurry, Ted remained focused on the inventions supporting the industry. At his backyard laboratory in Auburn, New York, he further perfected the AEO light and discovered a way to reduce background noise on film.

To achieve the latter, Ted borrowed an idea from transoceanic telephones. He saw how sending stations operated automatically one end station at a time. Ted set up a similar circuit with his recording lights. In the relay, words and desired sounds would trigger the lights to achieve degrees of brilliance. At other times, lights would dim, preventing the taping of audible disturbances. Ted's device opened new possibilities for settings and locations, and the movie industry would soon utilize its creation.

Filmed on Location

For Christmas 1928, Fox-Case released the first sound picture recorded outdoors. Crews shot *In Old Arizona* at many sites, including the Mojave Desert and the San Fernando Valley.

The realistic sounds surprised some spectators. Viewers were prepared to hear people speak, but they found other sounds quite startling: the whistle of a passing train, a rooster crowing in the distance, the crackling of bacon on the campfire skillet, the clip-clop of horses' hooves that actually receded as the rider galloped away from the camera!

An Academy Award went to Warner Baxter, the film's star who portrayed the Cisco Kid, and the landmark film received excellent reviews. Critics agreed that the novelty of filming on location was a needed boost to the industry. The clearness and naturalness of the sounds attracted wide attention. Ted's technology had given a fresh dimension to the Western movie, when critics thought it to be at the end of its trail.

Mr. William Fox

William Fox was an entrepreneur and a visionary in the movie industry. When the Big Five Studios (Metro-Goldwyn-Mayer, First National, Paramount, Universal, and Producers' Distributing Corporation) assumed a wait-and-see approach to talking movies, Mr. Fox gambled heavily that the days of silent movies were numbered.

What Mr. Fox did not know, however, was whether Ted's sound-on-film system or another system, like Vitaphone's sound-on-disk, would win out. To protect his interests, he hedged his bets.

Mr. Fox arranged a meeting with Warner Brothers and Western Electric, the company supplying amplifiers for the Vitaphone system. The men signed a contract to protect their business interests should one of their sound processes be victorious over the other. Each agreed to install both types of sound systems, sound-on-film and sound-on-disk, in their theaters. Furthermore, each granted the other a full legal license to use the sound system over which they had control.

Mr. Fox made similar deals with other production companies. The business arrangements worked to his advantage. The Fox empire was on the rise.

Ted, nonetheless, never knew of Mr. Fox's connections with other companies, even though their agreement stated that any change in contract required permission in writing. Ted never gave permission for other studios to use his inventions. Each of Mr. Fox's deals, therefore, lessened the value of the patents Ted had entrusted to the Fox-Case Corporation.

Outfoxed

On at least one occasion, Mr. Fox was outfoxed.

In fall 1928, Universal Studios asked to borrow one of the trucks used to shoot Movietone's newsreels. Universal's sound stages were not quite ready and the company said it needed equipment to shoot some tests.

Once behind closed doors, however, the studio quickly shot the first 100 percent talking motion picture with sound-on-film. The movie, *Melody of Love* starring Walter Pidgeon, was a huge success at the box office. Ablaze with its triumph, the studio continued using the equipment, adding sound sequences to three of its silent movies.

When Mr. Fox got wind of what had happened, he was outraged. Universal returned the newsreel truck, but the incident had a major impact on motion picture history. The unauthorized use of Movietone equipment further diminished the name of Ted Case as the creator of sound-on-film.

Sound-on-film Prevailed

By the end of the decade, there was an explosion of talking movie genres: musicals, Westerns, gangster movies, comedies, dramas and operettas. Stars like Myrna Loy, Maurice Chevalier, Loretta Young, Mary Pickford, Victor McLaglen, Frederic March, Douglas Fairbanks, Sophie Tucker, Stepin Fetchit, Fay Wray, Johnny Weissmuller, William Boyd, Clara Bow, Carole Lombard, Noah Beery, Beatrice Lillie, Zazu Pitts, William Powell, Hedda Hopper, Greta Garbo, John Gilbert, Jack Oakie, Rudy Vallee and Rin Tin Tin were becoming household names.

Fox Films was the first studio to announce it would cease production of silent movies. Columbia Pictures released a similar statement a few days later. One by one, motion picture companies declared that the silent movie had come to an end, and that they were now in the sole business of making talkies, talkies and more talkies.

By the end of 1929, sound-on-film was adopted by all but one movie studio.* Yet the various producers and movie engineers had a hard time explaining the origin of their sound technology. Strangely enough, theirs bore a striking resemblance to the synchronized system first invented in Auburn, New York, by Ted Case.

---

* *Warner Brothers discontinued using disk and converted to sound-on-film in 1930.*

This is a 1928 photograph of the Lafayette Theater in Batavia, New York. Notice the words Vitaphone and Movietone on the side marquee. William Fox and Jack Warner made a deal and installed both sound systems in their theaters. [Collection of the Case Research Lab Museum]

Tragic Times

On July 17, 1929, Mr. Fox was seriously injured in a car crash. During his convalescence, business rivals closed in on him. Western Electric, which had recently loaned Mr. Fox $15 million, said that Mr. Fox could pay up by handing over the patents for the sound-on-film process.

On September 20, 1929, Mr. Fox negotiated a deal with Ted. He offered to trade stocks in Fox Theater for Ted's share in their company. Ted accepted. The Fox-Case Corporation changed to the Fox Corporation, and Ted's sound-on-film patents became the legal property of Mr. Fox. Ted, however, would not be able to redeem the stocks until September 1930.

No one could predict the tragic times ahead. In October 1929, the United States stock market crashed. The worst economic collapse in the history of the modern industrial world found Mr. Fox's empire in a condition of extraordinary overexpansion. Exorbitant debts of more than $90 million forced Mr. Fox to sell out. The Fox name would continue in the movie industry, but the era of William Fox had ended.

☙

Remember Janie, that when I receive an idea I know that at least 200,000 people around the world receive it too but they brush it off.

Ted Case enjoyed his family and the natural beauty of Casowasco (Case + Owasco Lake). In the photograph with Ted are his mother, Eva, and his daughters Barbara (left) and Janie. [Collection of the Case Research Lab Museum]

X. The After Years:

(1930 – 1944)

Chimneys

Ted was happy to be back in Auburn. On South Street, he built a magnificent Tudor style mansion with libraries, ballrooms, playrooms for the children, and a 40-foot indoor swimming pool complete with below-surface lighting. The structure was named The Chimneys as a fireplace was built in nearly all of the 40 rooms.

The Case mansion was a gathering place for important meetings and for elaborate parties. Ted and his wife welcomed many well-known people, including the popular star of silent movies, Charles Chaplin, and the handsome cowboy of talking Westerns, John Wayne.

The mansion welcomed many Auburnians as well. Ted loved to play bridge. He also belonged to the local stamp club. After meetings, he would invite fellow club members in for a chat. One friend was a contractor from nearby Moravia named Albert Parker. Upon returning home, Mr. Parker would tell about Italian marble mantels, a gold-papered room and gold faucets. But his jovial host's spirited conversations and comfortable camaraderie provided his fondest memories of the evenings.

Casowasco

Ted also loved relaxing with his family at Casowasco, their summer home on Owasco Lake. He was an avid athlete and taught his children, now numbering four, to enjoy a variety of water and land sports.

Ted's daughter, Jane, recalled that Casowasco was truly beautiful and magical. "There were friends and family running and playing and laughing and eating delicious meals and enjoying each other, and music and enjoyment of the natural beauty we all lived in."

Visitors were always welcome. Walt Disney and Jack Warner would take a train to Auburn and then drive south toward Moravia. Other men would reach Casowasco after walking miles on the train tracks that ran along the shoreline. These men came to talk to Ted about their thoughts and dreams. Ted always listened. He advised the men to start acting upon their ideas and to begin right in their own back yards. He encouraged them with a strong belief he often shared with daughter Jane:

Remember Janie, that when I receive an idea I know that at least 200,000 people around the world receive it too but they brush it off. However, I also know that at least six other people somewhere in the world are taking the idea to heart and are working on it.

Ted also needed time to be alone at Casowasco. His typical attire was a tweed jacket, white flannel pants, white shirt and tie, and a dark red carnation. In the presence of family he would sometimes walk down to the dock, remove his jacket, tie, watch and shoes, go into the water and start floating. Ted later explained to Jane that if he went inside to get his bathing suit, someone from the large staff would stop him for one reason or another and he would never get back to the relaxing water ... and to *his* ideas.

Compassion

Ted showed compassion for people. After parties, he would give away gallons of ice cream to George Buggy, his chauffeur, who shared the treat with his family and neighbors on nearby Woodlawn Avenue.

He continued to invite groups of Auburnians to the carriage house. The *Auburn Citizen* (February 3, 1928) reported on an evening when 20 members of a Bible class from the Second Presbyterian Church were guests at the Case Research Laboratory on West Genesee Street:

> The young men saw one of the greatest inventions of the age.
> Theodore W. Case freely explained how talking movies were produced. This followed with a demonstration such as but few boys in the world today have had an opportunity to view. They saw the great U. S. Government battleship *Texas* steam into Havana Harbor bearing the president of the United States, heard the booming of the cannon ... and the wild acclaim and shouting of the immense throng of people

assembled to meet the distinguished guests arriving to attend the Pan American conference.

Other favorite talking movies of world renown were Charles Lindbergh's great reception at Washington. The boys ... heard Lindy make his great speech of appreciation to the American people

The boys were slow to leave the buildings and it is doubtful if they slept much last night after observing some of the wonders of the age and visiting such a personally conducted tour as they were privileged to enjoy last night at the Case laboratories.

Ted's concern for those less fortunate also continued. During the Great Depression, he visited friends in the area who were farmers. He wanted to see how they were getting along. When Ted sensed an economic struggle, he quietly arranged to pay the mortgage until better times returned.

The Last Years

The Depression eventually took its toll on Ted. In 1936, due to mounting expenses on his several properties, he arranged to give the Willard Estate to the city of Auburn for the sum of $5 and a box of cigars. He asked that the buildings and lab be a memorial to his father and a museum for Movietone. In 1938 Ted donated the South Street mansion to Auburn and took up residence at Casowasco and in New York City.

Ted remained active in the last years of his life. A 1937 Yale alumni publication included his remarks on healing with high frequency sound waves:

> I have always been most interested in scientific subjects, especially the branches having to do with light and electrons and more lately with supersonics or inaudible sound waves .... I believe that some day there will be a new and large branch of medicine which we might call Sound Therapy.
>
> That is what I am interested in at present. Have designed some new apparatus and am now working on the effect of high frequency sound vibrations directly applied locally to glands, and muscles of a living body.

Ted pursued other interests as well. In 1940, he held the position of vice president of the Pan American Trade Committee, whose headquarters were in New York City. He worked to promote good will between South America and the United States.

Affiliations

Ted served as vice president and a trustee of the Case Memorial Library and a trustee of the Cayuga Museum of History and Art. He was a director of the Auburn Trust Company and the General Banknote Engineering Corporation of New York.

Ted was affiliated with St. Peter's Episcopal Church in Auburn and St. James Church in New York City. He belonged to the Metropolitan, Yale and River clubs of New York, the Owasco Country and Yacht clubs of Auburn, the Skaneateles [N.Y.] Country Club, the Cayuga County Sportsman, Finger Lakes, and Automobile associations.

He was an active member in the American Association for the Advancement of Science, the American Physical Society, the American Electro Chemical Society, the Optical Society of America, the Royal Society of Arts (London), and the New York Electrical Society.

❧

Ted acquired 62 United States patents that shaped the development of talking pictures.

In this photograph, Ted Case is relaxing on the porch of his summer home at Casowasco. [Collection of the Case Research Lab Museum]

XI. Reflections

Lost in the Rush

Ted Case died in the spring of 1944 at the age of 55. He was a devoted father and husband. He was himself — whether dining with the Rockefellers or lunching with his chauffeur. He had spiritual awareness, tremendous concentration, and a great sense of humor. He had a strong yet humble character and heroic scientific steadfastness.

Unlike other scientists who tried to assemble a sound-on-film system, Ted worked in the application of basic sciences and created technical tools that were available nowhere else. In a backyard laboratory in Auburn, New York, he discovered a way for sound and action to be synchronized on moving film and then replayed as a natural illusion.

Ted acquired 62 United States patents that shaped the development of talking pictures. Yet recognition for his work was lost in the rush to revolutionize motion pictures. *

* *In 1946, through the efforts of Earl I. Sponable, the Society of Motion Picture Engineers added the name of Theodore W. Case to its "Honor Roll."*

One day his daughter came home all upset. She had just read an article in the *Saturday Evening Post* that gave credit for her father's creation to someone else. Ted read the article and turned to his daughter. "It's all right, Janie," he told her. "It doesn't matter what others think. I know what I did."

అ

Case Research Lab Museum

The Case Research Lab Museum opened in 1994 as a tribute to this great man of science.* Visitors can view some of Ted Case's early sound films. These films include a 1924 recording of Ted reciting Lincoln's Gettysburg Address, the 1925 sound test of Gus Visser and his singing duck, a 1926 interview with prison reformer, Thomas Mott Osborne, and the 1929 Auburn Prison riot.

One can also see Ted's chemistry lab, research apparatus, infrared signaling equipment, darkroom, recording studio and the sound film projector that set recording standards still used today.

ॐ

* *Throughout the year, thousands of visitors are welcomed at the Case Research Lab Museum and the Cayuga Museum of History and Art.*

These photographs, taken around 1920, show the Case Research Laboratory in Auburn, New York. The lab is the birthplace of two famous inventions: the invisible signaling system used in World War I and the process of sound-on-film. [Collection of the Case Research Lab Museum]

Dates in the life of Ted Case:

December 12, 1888 - Ted Case is born

1912 - Graduates from Yale University

1916 - Invents Thalofide Cell

1917 - Invents World War I secret
communication system

1918 - Marries Alice Gertrude Eldred

1922 - Uses AEO light to record sound on
moving film

1923 - Forms partnership with Dr. Lee DeForest

1924 - Makes first sound film of an American
president

1925 - Ends business relationship with Dr.
DeForest

1926 - Begins Fox-Case Corporation with
William Fox
Builds first sound stage studio in New
York City

1927 - Movietone films first sound newsreel outside studio walls
- Movietone films Charles Lindbergh's transatlantic takeoff
- Movietone films people and events worldwide

1928 - Releases first sound picture recorded outside on location (*In Old Arizona*)

1929 - Ends Fox-Case Corporation
- The Great Depression begins

1937 - Explores the use of sound vibrations in healing

1940 - Serves as vice president of Pan American Trade Committee

May 13, 1944 - Ted Case dies

This photograph shows one of the sound motion picture cameras used by the Fox-Case Corporation between 1926 and 1929.
Now we're talking! [Collection of the Case Research Lab Museum]

Printed in the United States
1446000005B/139-1110